習水古茶樹

ANCIENT TEA TREES BY XI RIVER

习水古茶树

ANCIENT TEA TREES BY XI RIVER

主编 谭智勇

人民出版社

Prologue

Xishui County, the native place of Xi State, is a red city in an oasis. Amid wild mountains grow vigorous and verdant ancient tea trees, which are the physical and cultural heritage passed down to descendents by Xi ancestors. For thousands of years, the ancient courier station in mountains has always been connected to the far-reaching Horse and Tea Trade Route, green leaves of dense forest have been tied up with the remote Ancient Silk Road, and rugged mountain roads under ancient trees have been left with footprints of the Red Army towards the Tiananmen Square. This is the ancient tea tree in Xishui and the elderly in Xibu (Xishui's ancient name, which dates back to the Qin and Han Dynasties), and this is the ancient tea tree in hometown and a green leaf in primitive forest,Telling the story of the hometown, the red gene of Xishui goes down generation by generation.

To exhibit Xishui's time-honored ancient tea culture and rich ancient tea tree resources in a comprehensive and visualized manner, unveil the mystery of the county's ancient tea tree, and make humble contribution to its poverty alleviation and economic and social development, Guizhou Xiguo Guli Tea Industry Development Co., Ltd. organized nostalgic photographers to launch the large-scale "Xishui Ancient Tea Tree" photography activity. The shooting activity lasts 70 days from August 15 to October 31, 2017. The photographers stepped into Xishui's deep mountains and forests with an altitude above 1,000 meters, and shot almost 600 ancient tea trees aged more than 300 years.

The vigorous ancient tea trees standing firm are rooted in fertile ground of the native place of Xi State, the running mountain ridges stretching thousands of miles away. The rainfall from the heaven and water at tree root converge to form brooks and further flow into the Chishui River where the Kweichow Maotai was originated from. The delicacy of the state liquor lies in the river from which spirit is fetched for brewing, while the river for brewing liquor stems from the tree of ancient tea, so the Emperor Wu of Han appreciated it as "so luscious".

Various clusters of the weathered ancient tea trees have always been there in the mountains of Xibu forests regardless of hardships and holding fast to their belief. The steep mountains and dense tea set off the winding and rugged ancient route. In the afternoon of January 24, 1935 when the Red Army setting out from Zunyi and led by Chairman Mao took their path on a tea mountain via Xishui Houtan tea garden village and stopped on the way for a rest under the thousand-year ancient tea tree in front of villager Mu Xiancai's home, the countryman brewed a big pot of aged tea to quench the soldiers' thirst. Chairman Mao, standing beneath an ancient tea tree, looked into the distance over steaming hot intense tea…

The ancient tea tree forests which have gone through ups and downs, multiply in the embrace of the hometown, leaving descendents fragrance all over the fields. Year over year, in silence, they remains true to their original aspiration and guards their hometown. They build a green wall to shelter villagers in windy days; they nurture bamboo forest in mountains to give villagers hope in rainy days. The vigorous trunk symbolizes the unyielding backbone of the elders, the luxuriant branches and leaves present the smiling faces of their offspring, and green shoots are the blessing of their descendents.

Each time we press the shutter, we were experiencing a sense of shock; standing before each ancient tea tree, we were receiving pure baptism. Only in the boundless tea mountains and only when hugging each ancient tea tree aged hundreds or even thousands of years can we know what is history, what behavior can be deemed as steadfast, what the saying that one generation plants the trees in whose shade another generation rests implies, and what hardships and striving should be like, thus respect, admiring and awe rising up.

As the Xishui people and descendents of the native place of Xi State, we extend heartfelt gratitude to the Xi ancestors and to the predecessors of Xibu for they leave us with the overspread ancient tea trees and the ancient tea garden featuring green hills and clear waters with wisdom, efforts and life from generation to generation.

General Secretary Xi Jinping has put it well, "green hills and clear water are gold and silver."

Xishui ancient tea tree is an evergreen in our life.

This is the first picture scroll of ancient tea tree in China to pay attribute to history as well as to the new era.

This is a profound respect paid by Xishui offspring to the future and to the whole world.

By Tan Zhiyong
January 1, 2018

序

  习水，鳛国故里，绿洲红城。莽莽群山之中生长着苍劲葱郁的古茶树，那是习氏先民留给后人的物质和文化遗产。千百年来，大山里的古驿栈总是连着远去的茶马古道，密林中的绿叶一直系着遥远的古丝绸之路，古树下的崎岖山路留下了红军走向天安门广场的足迹。这就是习水的古茶树，如同鳛部的老人，这就是家乡的古茶树，原始森林中的一片绿叶，讲述着家乡悠久的历史，传承着习水红色的基因。

  为了全面直观展现习水历史悠久的古茶文化和丰富的古茶树资源，揭开习水古茶树神秘的面纱，为习水扶贫攻坚和经济社会发展贡献绵薄之力，贵州鳛国故里茶业发展有限公司组织了热爱家乡的摄影人进行这次《鳛水古茶树》大型摄影活动。拍摄活动从 2017 年 8 月 15 日至 10 月 31 日，历时 70 天。摄影人走进习水县海拔 1000 米以上的深山老林，拍摄了近 600 余棵树龄在 300 年以上的古茶树。

  一棵棵苍劲挺立的古茶树，扎根在鳛国故里的沃土之中，山峦叠翠，绵延千里。天降甘霖，树根之水，潺潺流出，集成小溪，汇入赤水河，酿得美酒茅台。国酒之好在于美酒之河，美酒之河源于古茶之树，故汉武帝不得不感叹"甘美之"。

  一丛丛苍茫岁月的古茶树，守候在鳛部丛林的群山之中，不畏艰辛，信念坚定。山险茶密，掩映古道，弯弯曲曲，坎坷不平。1935 年 1 月 24 日下午，毛泽东主席率领从遵义出发的红军，沿着茶山小路途经习水吼滩茶园村，在村民穆贤才家门前那棵千年古茶树下歇脚，老乡熬了一大锅老茶给红军解渴。毛主席站在古茶树下，喝着热气腾腾的酽茶，遥望着远方……

  一片片沧海桑田的古茶树，繁衍在故乡的怀抱之中，留给后人，遍野清香。年复一年，默默无语，不忘初心，守候故土。风来了，搭建绿墙，给村民们挡风遮寒；雨落了，润育山箐，给村民们带来希望。苍劲的树干是老人不倔的脊梁，繁茂的枝叶是子孙们的笑脸，青青的嫩芽是后人们的福份。

  每按动一次快门，我们的心灵都感受到一次强烈的震撼；每来到一棵古茶树前，我们的心灵都经受到一次纯净的洗礼。只有置身在这无边无际的茶山中，只有拥抱着一棵棵百年、千年的古茶树，我们才知道什么叫历史，才清楚什么叫坚守，才明白什么叫前人栽树后人乘凉，才能体会到习氏先民们的艰辛和奋斗，从而产生对先辈们的敬重、敬仰和敬畏。

  作为习水人，作为鳛国故里的后代，我们深深感谢习氏的先民们，感谢鳛部的先辈们，是他们一代又一代用智慧、汗水和生命给习水人民留下了这漫山遍野的古茶树，给我们留下了这山青水秀的古茶园。

  习近平总书记说得好："绿水青山，就是金山银山"。

  习水古茶树，是我们生命中的长青树。

  这是中国第一部古茶树画卷，献给历史，献给新时代。

  这是习水儿女一片深深敬意，献给未来，献给全世界。

谭智勇

2018 年 1 月 1 日

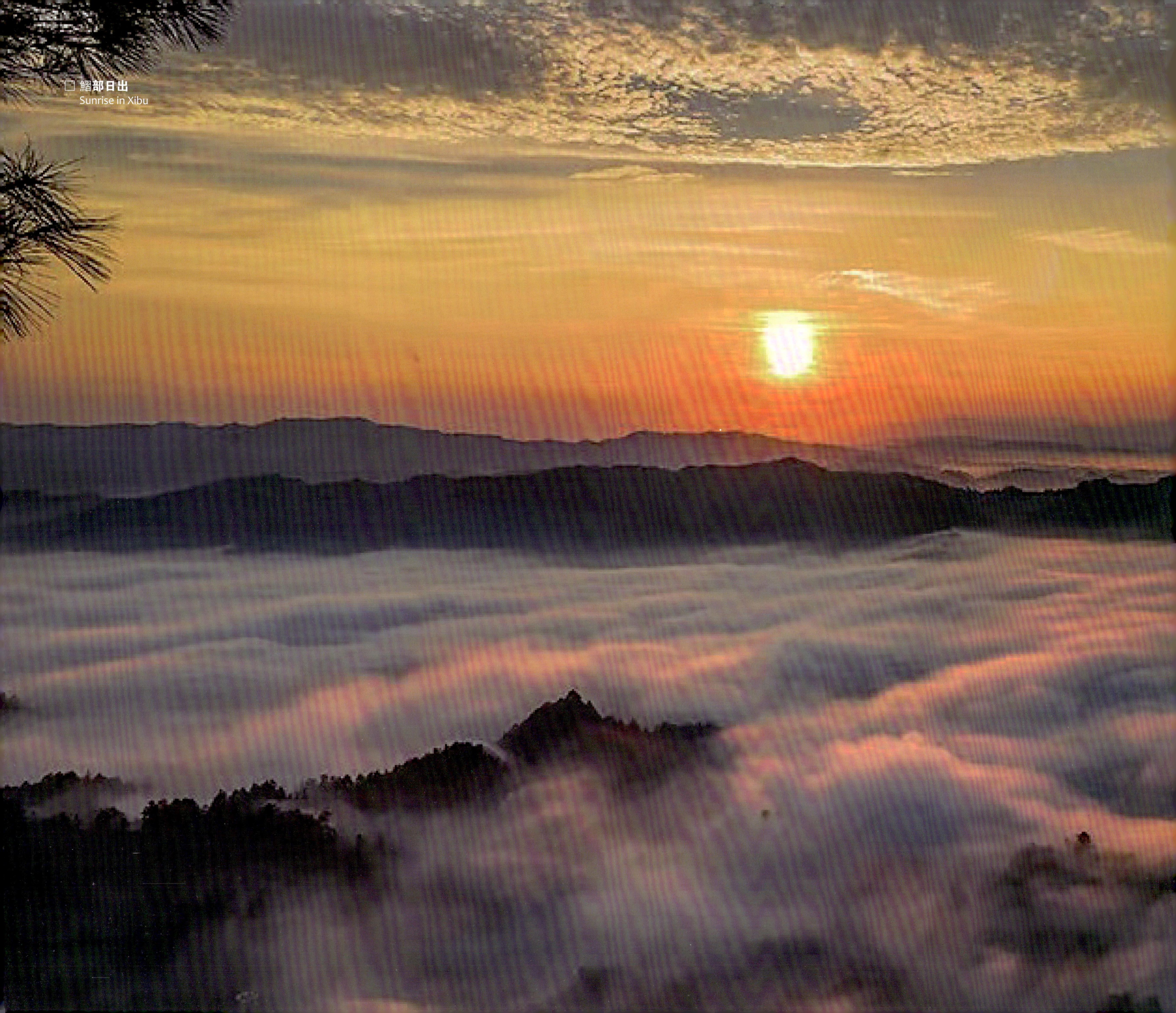

□ 鳛部日出
Sunrise in Xibu

□ 习水群山
Mountains in Xishui

□ 习水古茶树林
Ancient tea forest in Xishui

□ 古茶树下的村寨—桑木镇大山村
Dashan Village, Sangmu Town- a Village of Ancient Tea Trees

□ 千年古茶树良村镇羊化村
Yanghua Village, Liangcun Town, with thousand-year-old ancient tea trees
□ 树高 8m 茎围 200cm 树龄 1400 年
With a height of 8 meters, a circumference of 200cm and an age of 1400 years

古树新芽
New sprouts from ancient trees

□ 杉王街道羊九村茶田组（四沟头）
(Sigoutou) Chatian Group, Yangjiu Village, Shanwang Sub-District
□ 树高 10m 胸围 160cm 树龄 1000 年
With a height of 10 meters, a circumference of 160cm and an age of 1000 years

☐ **仙源镇金桥村丫口组**
Yakou Group, Jinqiao Village, Xianyuan Town
☐ 树高 11m 胸围 150cm 树龄 900 年
With a height of 11 meters, a circumference of 150cm and an age of 900 years

☐ **仙源镇金桥村街上组（箐角）**
(Qingjiao)Jieshang Group, Jinqiao Village, Xianyuan Town
☐ 树高 8m 胸围 100cm 树龄 400 年
With a height of 8 meters, a circumference of 100cm and an age of 400 years

□ 仙源镇毛坪村七阳平组（七阳平）
(Qiyangping) Qingyangping Group, Maoping Village, Xianyuan Town
□ 树高 9m　胸围 190cm　树龄 1300 年
With a height of 9 meters, a circumference of 190cm and an age of 1300 years

□ **仙源镇金桥村丫口组**
Yakou Group, Jinqiao Village, Xianyuan Town
□ 树高 11m 胸围 100cm 树龄 400 年
With a height of 11 meters, a circumference of 100cm and an age of 400 years

□ **仙源镇羊九村消抗坪组（消抗坪）**
(Xiaokangping) Xiaokangping Group, Yangjiu Village, Xianyuan Town
□ 树高 7m 胸围 95cm 树龄 300 年
With a height of 7 meters, a circumference of 95cm and an age of 300 years

仙源镇毛坪村七阳平组（七阳平）
(Qiyangping) Qingyangping Group, Maoping Village, Xianyuan Town
树高 9m 胸围 190cm 树龄 1300 年
With a height of 9 meters, a circumference of 190cm and an age of 1300 years

仙源镇金桥村街上组（箐角）
(Qingjiao) Jieshang Group, Jinqiao Village, Xianyuan Town
树高 8m 胸围 90cm 树龄 300 年
With a height of 8 meters, a circumference of 90cm and an age of 300 years

□ 仙源镇仙源村金桥组
Jinqiao Group, Xianyuan Village, Xianyuan Town

□ 树高 7m 胸围 120cm 树龄 600 年
With a height of 7 meters, a circumference of 120cm and an age of 600 years

□ 仙源镇仙源村金桥组
Jinqiao Group, Xianyuan Village, Xianyuan Town

□ 树高 6m 胸围 90cm 树龄 300 年
With a height of 6 meters, a circumference of 90cm and an age of 300 years

□ 仙源镇金桥村大坪组
Daping Group, Jinqiao Village, Xianyuan Town
□ 树高 8m 胸围 120cm 树龄 600 年
With a height of 8 meters, a circumference of 120cm and an age of 600 years

□ **仙源镇金桥村大坪组**
Daping Group, Jinqiao Village, Xianyuan Town

□ 树高 8m 胸围 120cm 树龄 600 年
With a height of 8 meters, a circumference of 120cm and an age of 600 years

□ **仙源镇金桥村大坪组**
ODaping Group, Jinqiao Village, Xianyuan Town

□ 树高 8m 胸围 130cm 树龄 700 年
With a height of 8 meters, a circumference of 130cm and an age of 700 years

□ 仙源镇金桥村古茶树与蓝天
Ancient tea trees and a blue sky in Jinqiao Village, Xianyuan Town

□ 仙源镇金桥村街上组（箐角）
(Qingjiao)Jieshang Group, Jinqiao Village, Xianyuan Town

□ 树高 8m　胸围 90cm　树龄 300 年
With a height of 8 meters, a circumference of 90cm and an age of 300 years

□ 仙源镇金桥村街上组（菁角）
(Qingjiao)Jieshang Group, Jinqiao Village, Xianyuan Town

□ 树高 7m 约冠 110cm 树龄 300 年
With a height of 7 meters, a circumference of 110cm and an age of 500 years

□ 仙源镇金桥村街上组（箐角）
(Qingjiao)Jieshang Group, Jinqiao Village, Xianyuan Town

□ 树高 8m 胸围 90cm 树龄 300 年
With a height of 8 meters, a circumference of 90cm and an age of 300 years

□ 仙源镇金桥村街上组（箐角）
(Qingjiao)Jieshang Group, Jinqiao Village, Xianyuan Town

□ 树高 8m 胸围 120cm 树龄 600 年
With a height of 8 meters, a circumference of 120cm and an age of 600 years

□ **仙源镇毛坪村河坝组（河坝）**
(Heba)Heba Group, Maoping Village, Xianyuan Town
□ 树高 6m　胸围 100　树龄 400 年
With a height of 6 meters, a circumference of 100cm and an age of 400 years

□ 仙源镇毛坪村河坝组（河坝）
(Heba)Heba Group, Maoping Village, Xianyuan Town

□ 树高 6m 胸围 80 树龄 300 年
With a height of 6 meters, a circumference of 80cm and an age of 300 years

□ 仙源镇羊九村四岗组（四岗）
(Sigang) Sigang Group, Yangjiu Village, Xianyuan Town

□ 树高 6m 胸围 80cm 树龄 300 年
With a height of 6 meters, a circumference of 80cm and an age of 300 years

□ 仙源镇大獐村马家垭口组（马家垭口）
(Majiayakou) Majiayakou Group, Dazhang Village, Xianyuan Town
□ 树高 7m 胸围 80cm 树龄 300 年
With a height of 7 meters, a circumference of 80cm and an age of 300 years

□ 仙源镇羊九村四岗组（四岗）
(Sigang) Sigang Group, Yangjiu Village, Xianyuan Town
□ 树高 7m 胸围 80cm 树龄 300 年
With a height of 7 meters, a circumference of 80cm and an age of 300 years

□ 仙源镇金桥村南山组（猴子岩）
(Monkey Rocks) Nanshan Group, Jinqiao Village, Xianyuan Town

□ 树高 7m　胸围 110cm　树龄 500 年
With a height of 7 meters, a circumference of 110cm and an age of 500 years

□ 仙源镇金桥村大坪组
Daping Group, Jinqiao Village, Xianyuan Town

□ 仙源镇金桥村街上组（箐角）
(Qingjiao)Jieshang Group, Jinqiao Village, Xianyuan Town

□ 仙源镇羊九村四岗组（四岗）
(Sigang) Sigang Group, Yangjiu Village, Xianyuan Town
□ 树高 9m　胸围 160cm 1000 年
With a height of 9 meters, a circumference of 160cm and an age of 1000 years

□ 仙源镇羊九村四岗组（四岗）
(Sigang) Sigang Group, Yangjiu Village, Xianyuan Town
□ 树高 9m　胸围 160cm 1000 年
With a height of 9 meters, a circumference of 160cm and an age of 1000 years

□ 仙源镇羊九村四岗组（四岗）
(Sigang) Sigang Group, Yangjiu Village, Xianyuan Town
□ 树高 9m 胸围 160cm 树龄 1000 年
With a height of 9 meters, a circumference of 160cm and an age of 1000 years

□ 仙源镇羊九村四岗组（四岗）
(Sigang) Sigang Group, Yangjiu Village, Xianyuan Town
□ 树高 9m 胸围 160cm 树龄 1000 年
With a height of 9 meters, a circumference of 160cm and an age of 1000 years

□ 仙源镇大獐村马家垭口组（马家垭口）
(Majiayakou) Majiayakou Group, Dazhang Village, Xianyuan Town
□ 树高 5m 胸围 70cm 树龄 250 年
With a height of 5 meters, a circumference of 70cm and an age of 250 years

□ 仙源镇羊九村白水井组（坟林头）
(Fenlintou)Baishuijing Group, Yangjiu Village, Xianyuan Town
□ 树高 9m 胸围 240cm 树龄 2200 年
With a height of 9 meters, a circumference of 240cm and an age of 2200 years

仙源镇大漳村马家垭口组（马家垭口）
(Majiayakou) Majiayakou Group, Dazhang Village, Xianyuan Town
树高 6m 胸围 95cm 树龄 300
With a height of 6 meters, a circumference of 95cm and an age of 300 years

□ 仙源镇大獐村陆井湾组（清龙沟）
(Qinglonggou) Lujingwan Group, Dazhang Village, Xianyuan Town
□ 树高 6m 胸围 80cm 树龄 300 年
With a height of 6 meters, a circumference of 80cm and an age of 300 years

□ 仙源镇大獐村陆井湾组（清龙沟）
(Qinglonggou) Lujingwan Group, Dazhang Village, Xianyuan Town

□ 树高 6m 胸围 70cm 树龄 250 年
With a height of 6 meters, a circumference of 70cm and an age of 250 years

□ 仙源镇大獐村马家垭口组（马家垭口）
(Majiayakou) Majiayakou Group, Dazhang Village, Xianyuan Town

□ 树高 8m 胸围 130cm 树龄 700 年
With a height of 8 meters, a circumference of 130cm and an age of 700 years

仙源镇大獐村陆井湾组（文家园子）
(Wenjiayuanzi) Lujingwan Group, Dazhang Village, Xianyuan Town

树高 10m 胸围 130cm 树龄 700 年
With a height of 10 meters, a circumference of 130cm and an age of 700 years

仙源镇大獐村陆井湾驵（清龙沟）
(Qinglonggou) Lujingwan Group, Dazhang Village, Xianyuan Town

树高 8m 胸围 90cm 树龄 300 年
With a height of 8 meters, a circumference of 90cm and an age of 300 years

仙源镇大樟村马家垭口（马家垭口）
(Majiayakou) Majiayakou Group, Dazhang Village, Xianyuan Town
树高 8m　胸围 90cm　树龄 300 年
With a height of 8 meters, a circumference of 90cm and an age of 300 years

□ 程寨镇印江村大坪组（罗家湾子）
(Luojiawanzi) Daping Group, Yinjiang Village, Chengzhai Town
□ 树高 10m 胸围 130cm 树龄 700 年
With a height of 10 meters, a circumference of 130cm and an age of 700 years

□ 程寨镇印江村大坪组（罗家湾子）
(Luojiawanzi) Daping Group, Yinjiang Village, Chengzhai Town
□ 树高 11m 胸围 150cm 树龄 900 年
With a height of 11 meters, a circumference of 150cm and an age of 900 years

□ 程寨镇印江村罗家组（罗家湾子）
(Luojiawanzi) Baojia Group, Yinjiang Village, Chengzhai Town

□ 树高 8m　胸径 120cm1　树龄 600 年
With a height of 8 meters, a circumference of 120cm and an age of 600 years

□ 程寨镇印江村大坪组（罗家湾子）
(Luojiawanzi) Daping Group, Yinjiang Village, Chengzhai Town
□ 树高 7m　胸围 120cm　树龄 600 年
With a height of 7 meters, a circumference of 120cm and an age of 600 years

□ 程寨镇印江村大坪组（罗家湾子）
(Luojiawanzi) Daping Group, Yinjiang Village, Chengzhai Town
□ 树高 8m　胸围 120cm　树龄 600 年
With a height of 8 meters, a circumference of 120cm and an age of 600 years

程寨镇印江村大坪组（罗家湾子）
(Luojiawanzi) Daping Group, Yinjiang Village, Chengzhai Town
树高 7m，约围 130cm，树龄 700 年
With a height of 7 meters, a circumference of 130cm and an age of 700 years

□ **程寨镇印江村大坪组（罗家湾子）**
(Luojiawanzi) Daping Group, Yinjiang Village, Chengzhai Town
□ 树高 8m　胸围 125cm　树龄 600 年
With a height of 8 meters, a circumference of 125cm and an age of 600 years

程寨镇印江村大坪组（罗家湾子）古茶树林
Ancient tea forest, (Luojiawanzi) Daping Group, Yinjiang Village, Chengzhai Town

□ 程寨镇印江村大坪组（罗家湾子）
(Luojiawanzi) Daping Group, Yinjiang Village, Chengzhai Town
□ 树高 9m 胸围 135cm 树龄 700 年
With a height of 9 meters, a circumference of 135cm and an age of 700 years

□ **程寨镇印江村大坪组（罗家湾子）**
(Luojiawanzi) Daping Group, Yinjiang Village, Chengzhai Town
□ 树高 9m 胸围 130cm 树龄 700 年
With a height of 9 meters, a circumference of 130cm and an age of 700 years

□ **古茶树枝**
Ancient tea tree branches

程寨镇印江村大坪组（罗家湾子）
(Luojiawanzi) Daping Group, Yinjiang Village, Chengzhai Town

树高 11m　胸围 148cm　树龄 900 年
With a height of 11 meters, a circumference of 148cm and an age of 900 years

□ **程寨镇印江村大坪组（罗家湾子）**
(Luojiawanzi) Daping Group, Yinjiang Village, Chengzhai Town
□ 树高 11m　胸围 150cm　树龄 900 年
With a height of 11 meters, a circumference of 150cm and an age of 900 years

□ 二里镇星光村胡家组（黄泥塆）
(Huangniwan) Hujia Group, Xingguang Village, Erli Town
□ 树高 4m　胸围 88cm　树龄 300 年
With a height of 4 meters, a circumference of 88cm and an age of 300 years

□ 二里镇星光村黄家组（马扶祠）
(Mafu Temple) Huangjia Group, Xingguang Village, Erli Town
□ 树高 6m 胸围 73cm 树龄 260 年
With a height of 6 meters, a circumference of 73cm and an age of 26 years

□ 民化镇三元村大岩组（兴福寺）
(Xingfu Temple) Dayan Group, Sanyuan Village, Minhua Town
□ 树高 10m　胸围 85cm　树龄 300 年
With a height of 10 meters, a circumference of 85cm and an age of 300 years

□ 民化镇三元村大岩组（兴福寺）
(Xingfu Temple) Dayan Group, Sanyuan Village, Minhua Town
□ 树高 10m　胸围 120cm　树龄 600 年
With a height of 10 meters, a circumference of 120cm and an age of 600 years

□ 古树新芽
New sprouts from ancient trees

同民镇蔺江村五组（龙山）
(Longshan) Wuzu Group, Linjiang Village, Tongmin Town

□ 同民镇蔺江村五组（龙山）
(Longshan) Wuzu Group, Linjiang Village, Tongmin Town
□ 树高 4m 胸围 100cm 树龄 400 年
With a height of 4 meters, a circumference of 100cm and an age of 400 years

桑木镇大山村高杠子组
Gaogangzi Group, Dashan Village, Sangmu Town

☐ **桑木镇大山村高杠子组**
Gaogangzi Group, Dashan Village, Sangmu Town

☐ 树高 6m 胸围 120cm 树龄 600
With a height of 6 meters, a circumference of 120cm and an age of 600 years

☐ **桑木镇桐捲村弯头组（角耳山）**
(Jiao'er Mountain) Wantou Group, Tongquan Village, Sangmu Town

☐ 树高 6m 胸围 110cm 树龄 500 年
With a height of 6 meters, a circumference of 110cm and an age of 500 years

□ 桑木镇桐卷村坝上组（货家垭口）
(Huojiayakou)Bashang Group, Tongjuan Village, Sangmu Town
□ 树高 6m 胸围 115cm 树龄 500 年
With a height of 6 meters, a circumference of 115cm and an age of 500 years

□ **桑木镇大山村高杠子组**
Gaogangzi Group, Dashan Village, Sangmu Town

□ 树高 8m 胸围 140cm 树龄 800 年
With a height of 8 meters, a circumference of 140cm and an age of 800 years

□ **桑木镇大山村湾子组**
Wanzi Group, Dashan Village, Sangmu Town

□ 树高 4m 胸围 110cm 树龄 500 年
With a height of 4 meters, a circumference of 110cm and an age of 500 years

□ 桑木镇大山村农中组（林场）
(Forestry Station) Nongzhong Group, Dashan Village, Sangmu Town
□ 树高 9m 胸围 125cm 树龄 600 年
With a height of 9 meters, a circumference of 125cm and an age of 600 years

□ 桑木镇大山村湾子组
Wanzi Group, Dashan Village, Sangmu Town
□ 树高 6m 胸围 120cm 树龄 600 年
With a height of 6 meters, a circumference of 120cm and an age of 600 years

□ **桑木镇大山村湾子组**
Wanzi Group, Dashan Village, Sangmu Town
□ 树高 6m 胸围 100cm 树龄 400 年
With a height of 6 meters, a circumference of 100cm and an age of 400 years

□ **桑木镇大山村湾子组**
Wanzi Group, Dashan Village, Sangmu Town
□ 树高 7m 胸围 120cm 树龄 600 年
With a height of 7 meters, a circumference of 120cm and an age of 600 years

□ **桑木镇大山村农中组（道沟头）**
(Daogoutou) Nongzhong Group, Dashan Village, Sangmu Town
□ 树高 9m 胸围 70cm 树龄 250 年
With a height of 9 meters, a circumference of 70cm and an age of 250 years

□ **桑木镇大山村湾子组**
Wanzi Group, Dashan Village, Sangmu Town
□ 树高 5m 胸围 100cm 树龄 400 年
With a height of 5 meters, a circumference of 100cm and an age of 400 years

□ 桑木镇大山村农中组（林场）
(Forestry Station) Nongzhong Group, Dashan Village, Sangmu Town
□ 树高 10m 胸围 85cm 树龄 300 年
With a height of 10 meters, a circumference of 85cm and an age of 300 years

□ 桑木镇大山村农中组（林场）
(Forestry Station) Nongzhong Group, Dashan Village, Sangmu Town
□ 树高 5m 胸围 120cm 树龄 600 年
With a height of 5 meters, a circumference of 120cm and an age of 600 years

桑木镇大山村农中组
Nongzhong Group, Dashan Village, Sangmu Town
树高 8m 胸围 100cm 树龄 400 年
With a height of 8 meters, a circumference of 100cm and an age of 400 years

□ 桑木镇大山村农中组
Nongzhong Group, Dashan Village, Sangmu Town
树高 7m 胸围 100cm 树龄 400 年
With a height of 7 meters, a circumference of 100cm and an age of 400 years

□ **桑木镇大山村农中组（道沟头）**
(Daogoutou) Nongzhong Group, Dashan Village, Sangmu Town
□ 树高 7m 胸围 75cm 树龄 250 年
With a height of 7 meters, a circumference of 75cm and an age of 250 years

□ **桑木镇大山村湾子组（老房子）**
(Laofangzi) Wanzi Group, Dashan Village, Sangmu Town
□ 树高 9m 胸围 130cm 树龄 700 年
With a height of 8 meters, a circumference of 130cm and an age of 700 years

□ 桑木镇大山村农中组（林场）
(Forestry Station) Nongzhong Group, Dashan Village, Sangmu Town
□ 树高 10m 胸围 170cm 树龄 1100 年
With a height of 10 meters, a circumference of 170cm and an age of 1100 years

□ 桑木镇大山村湾子组
Wanzi Group, Dashan Village, Sangmu Town
□ 树高 10m 胸围 144cm 树龄 800 年
With a height of 10 meters, a circumference of 144cm and an age of 800 years

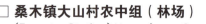

□ 桑木镇大山村湾子组
Wanzi Group, Dashan Village, Sangmu Town

□ 树高 9.5m 胸围 170cm 树龄 1100 年
With a height of 9.5 meters, a circumference of 170cm and an age of 1100 years

□ 桑木镇大山村农中组（道沟头）（兄弟树）
(Xiongdishu)(Daogoutou) Nongzhong Group, Dashan Village, Sangmu Town
□ 树高 13m 胸围 210cm 树龄 1400 年
With a height of 13 meters, a circumference of 210cm and an age of 1400 years

□ 桑木镇大山村农中组（道沟头）
(Daogoutou) Nongzhong Group, Dashan Village, Sangmu Town

□ 桑木镇大山村农中组（林场）
(Forestry Station) Nongzhong Group, Dashan Village, Sangmu Town
□ 树高 8m 胸围 130cm 树龄 700 年
With a height of 8 meters, a circumference of 130cm and an age of 700 years

桑木镇大山村湾子组
Wanzi Group, Dashan Village, Sangmu Town
树高 5m 胸围 140cm 树龄 800 年
With a height of 5 meters, a circumference of 140cm and an age of 800 years

□ 桑木镇大山村湾子组
Wanzi Group, Dashan Village, Sangmu Town
□ 树高 6m 胸围 150cm 树龄 900 年
With a height of 6 meters, a circumference of 150cm and an age of 900 years

□ **桑木镇大山村农中组**
Nongzhong Group, Dashan Village, Sangmu Town
□ 树高 10m　胸围 170cm　树龄 1100 年
With a height of 10 meters, a circumference of 170cm and an age of 1100 years

□ **桑木镇大山村农中组（林场）**
(Forestry Station) Nongzhong Group, Dashan Village, Sangmu Town

□ 树高 10m 胸围 144cm 树龄 800 年
With a height of 10 meters, a circumference of 144cm and an age of 800 years

□ **桑木镇大山村湾子组**
Wanzi Group, Dashan Village, Sangmu Town

□ 树高 9m 胸围 170cm 树龄 1100 年
With a height of 9 meters, a circumference of 170cm and an age of 1100 years

□ 桑木镇大山村农中组（道沟头）一片树林
A Forest, (Daogoutou) Nongzhong Group, Dashan Village, Sangmu Town

□ 桑木镇大山村湾子组
Wanzi Group, Dashan Village, Sangmu Town
□ 树高 8m 胸围 110cm 树龄 500 年
With a height of 8 meters, a circumference of 110cm and an age of 500 years

□ 桑木镇大山村高杠子组
Gaogangzi Group, Dashan Village, Sangmu Town

□ 桑木镇大山村农中组（道沟头）
(Daogoutou) Nongzhong Group, Dashan Village, Sangmu Town
□ 树高 8m 胸围 100cm 树龄 400 年
With a height of 8 meters, a circumference of 100cm and an age of 400 years

□ 桑木镇大山村农中组（林场）
(Forestry Station) Nongzhong Group, Dashan Village, Sangmu Town
□ 树高 9m 胸围 110cm 500 年
With a height of 9 meters, a circumference of 110cm and an age of 500 years

□ 桑木镇桐梾村湾头组（角耳山）
(Jiao'er Mountain) Wantou Group, Tongquan Village, Sangmu Town
□ 树高 7m 胸围 120cm 树龄 600 年
With a height of 7 meters, a circumference of 120cm and an age of 600 years

□ 桑木镇大山村农中组（道沟头）
(Daogoutou) Nongzhong Group, Dashan Village, Sangmu Town

□ 桑木镇大山村古茶树林
Ancient tea forest, Dashan Village, Sangmu Town

桑木镇大山村古茶树簇拥着的老民居
Time-honored residences amidst ancient tea trees, in Dashan Village, Sangmu Town

□ 杉王街道羊九村复兴组
Fuxing Group, Yangjiu Village, Shanwang Sub-district
□ 树高 8m 胸围 180cm 树龄 1200 年
With a height of 8 meters, a circumference of 180cm and an age of 1200 years

□ 杉王街道羊九村复兴组
Fuxing Group, Yangjiu Village, Shanwang Sub-district
□ 树高 9m 胸围 180cm 树龄 1200 年
With a height of 9 meters, a circumference of 180cm and an age of 1200 years

□ 杉王街道羊九村复兴组（大面上）
(Damianshang) Fuxing Group, Yangjiu Village, Shanwang Sub-district
□ 树高 8m 胸围 140cm 树龄 800 年
With a height of 8 meters, a circumference of 200cm and an age of 1400 years

□ 杉王街道羊九村复兴组
Fuxing Group, Yangjiu Village, Shanwang Sub-district
□ 树高 9m 胸围 180cm 树龄 1200 年
With a height of 9 meters, a circumference of 180cm and an age of 1200 years

□ 杉王街道羊九村玉米地里的古茶树
Ancient tea trees in the corn fields in Yangjiu Village, Shanwang Sub-district

□ 杉王街道天鹅池村大田坝组
Datianba Group, Tianechi Village, Shanwang Sub-district

□ 树高 6m 胸围 90cm 树龄 300 年
With a height of 8 meters, a circumference of 200cm and an age of 1400 years

□ 杉王街道羊九村福兴组（油葛坝）
(Yougeba) Fuxing Group, Yangjiu Village, Shanwang Sub-district

□ 树高 8m 胸围 150cm 树龄 900 年
With a height of 8 meters, a circumference of 200cm and an age of 1400 years

□ 杉王街道羊九村大杆烟组
Daganyan Group, Yangjiu Village, Shanwang Sub-district
□ 树高 10m 胸围 200cm 树龄 1400 年
With a height of 10 meters, a circumference of 200cm and an age of 1400 years

□ **杉王街道天鹅池硝厂组**
Xiaochang Group, Tianechi, Shanwang Sub-district
□ 树高 12m 胸围 150cm 树龄 900 年
With a height of 12 meters, a circumference of 150cm and an age of 900 years

☐ 杉王街道羊九村大杆烟组（红坝子）
(Hongbazi) Daganyan Group, Yangjiu Village, Shanwang Sub-district

☐ 树高 9m 胸围 95cm 树龄 300 年
With a height of 9 meters, a circumference of 95cm and an age of 300 years

☐ 杉王街道羊九村复兴组（四沟头）
(Sigoutou) Fuxing Group, Yangjiu Village, Shanwang Sub-district

☐ 树高 13m 胸围 170cm 树龄 1100 年
With a height of 13 meters, a circumference of 170cm and an age of 1100 years

□ **杉王街道羊九村复兴组**
Fuxing Group, Yangjiu Village, Shanwang Sub-district
□ 树高 8m 胸围 180cm 树龄 1200 年
With a height of 8 meters, a circumference of 180cm and an age of 1200 years

杉王街道羊九村复兴组
Fuxing Group, Yangjiu Village, Shanwang Sub-district
树高 8m 胸围 180cm 树龄 1200 年
With a height of 8 meters, a circumference of 180cm and an age of 1200 years

□ **杉王街道羊九村复兴组（四沟尖）**
(Sigoutou) Fuxing Group, Yangjiu Village, Shanwang Sub-district

□ 树高 13m　胸围 120cm　树龄 600 年
With a height of 13 meters, a circumference of 120cm and an age of 600 years

□ 杉王街道羊九村复兴组（大面上）
(Damianshang) Fuxing Group, Yangjiu Village, Shanwang Sub-district
□ 树高 7m 胸围 80cm 树龄 300 年
With a height of 7 meters, a circumference of 80cm and an age of 300 years

□ 杉王街道羊九村复兴组（复兴坝）
(Fuxingba) Fuxing Group, Yangjiu Village, Shanwang Sub-district
□ 树高 10m 胸围 100cm 树龄 400 年
With a height of 10 meters, a circumference of 100cm and an age of 400 years

良村镇羊化村三台山组
Santaishan Group, Yanghua Village, Liangcun Town

□ **良村镇良村村小沟组**
Xiaogou Group, Liangcun Village, Liangcun Town
□ 树高 8m 胸围 100cm 树龄 300 年
With a height of 8 meters, a circumference of 100cm and an age of 300 years

□ **良村镇茶园村茶村组（茶村）**
(Tea Village) Chacun Group, Chayuan Village, Liangcun Town
□ 树高 8.5m 胸围 142cm 树龄 800 年
With a height of 8.5 meters, a circumference of 142cm and an age of 800 years

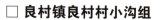

良村镇茶园村茶村组（茶村）
（Tea Village）Chaoun Group, Chayuan Village, Liangcun Town
树高 6.5 m 超围 150 cm 树龄 800 年
With a height of 6.5 meters, a circumference of 150cm and an age of 800 years

良村镇茶园村茶村组
Chacun Group, Chayuan Village, Liangcun Town
树高 11m　胸围 240cm　树龄 2200 年
With a height of 11 meters, a circumference of 240cm and an age of 2200 years

□ 良村镇茶园村菜村组
Chacun Group, Chayuan Village, Liangcun Town
□ 树高 8m 胸围 150cm 树龄 900 年
With a height of 8 meters, a circumference of 150cm and an age of 900 years

□ 良村镇羊化村半坎组
Bankan Group, Yanghua Village, Liangcun Town
□ 树高 6m 胸围 80cm 树龄 300 年
With a height of 6 meters, a circumference of 80cm and an age of 300 years

☐ **良村镇羊化村中心组（中岩沟）**
(Zhongyangou) Zhongxin Group, Yanghua Village, Liangcun Town

☐ 树高 6m 胸围 140cm 树龄 800 年
With a height of 6 meters, a circumference of 140cm and an age of 800 years

□ **良村镇茶园村茶村组（茶村）**
(Tea Village) Chacun Group, Chayuan Village, Liangcun Town

□ 树高 8.5m 胸围 140cm 树龄 800 年
With a height of 8.5 meters, a circumference of 140cm and an age of 800 years

□ **良村镇杨将村杨将组**
Yangjiang Group, Yangjiang Village, Liangcun Town
□ 树高 8m 胸围 90cm 树龄 300 年
With a height of 8 meters, a circumference of 90cm and an age of 300 years

□ 良村镇羊化村中心组（茶树湾）
(Chashuwan) Zhongxin Group, Yanghua Village, Liangcun Town

□ 树高 7m 胸围 100cm 树龄 400 年
With a height of 7 meters, a circumference of 100cm and an age of 400 years

□ 良村镇羊化村中心组（茶树湾）
(Chashuwan) Zhongxin Group, Yanghua Village, Liangcun Town

□ 树高 6m 胸围 120cm 树龄 600 年
With a height of 6 meters, a circumference of 120cm and an age of 600 years

良村頭杨将村
Yangjiang Village, Liangcun Tow

树高 9m 胸围 80cm 树龄 300 年
With a height of 9 meters, a circumference of 80cm and an age of 300 years

☐ **良村镇羊化村半坎组**
Bankan Group, Yanghua Village, Liangcun Town

☐ 树高 6m 胸围 79cm 树龄 300 年
With a height of 6 meters, a circumference of 79cm and an age of 300 years

☐ **良村镇羊将村羊将组**
Yangjiang Group, Yangjiang Village, Liangcun Town

☐ 树高 7m 胸围 110cm 树龄 500 年
With a height of 7 meters, a circumference of 110cm and an age of 500 years

□ 良村镇茶园村茶村组（茶村）
(Tea Village) Chacun Group, Chayuan Village, Liangcun Town

□ 树高 6m 胸围 83cm 树龄 300 年
With a height of 6 meters, a circumference of 83cm and an age of 300 years

□ 良村镇羊将村羊将组
Yangjiang Group, Yangjiang Village, Liangcun Town

□ 树高 7m 胸围 120cm 树龄 600 年
With a height of 7 meters, a circumference of 120cm and an age of 600 years

■ 良村镇羊化村中心组 茶林
Tea Forest，Zhongxin Group, Yanghua Village, Liangcun Town

良村镇杨将村杨将组
Yangjiang Group, Yangjiang Village, Liangcun Town
树高 9m　胸围 84cm　树龄 300 年
With a height of 9 meters, a circumference of 84cm and an age of 300 years

□ 良村镇杨将村杨将组　古树茶与狗
Yangjiang Group, Yangjiang Village, Liangcun Town Ancient tree tea and dogs

□ 良村镇杨将村杨将组
Yangjiang Group, Yangjiang Village, Liangcun Town

□ 树高 9m 胸围 80cm 树龄 300 年
With a height of 9 meters, a circumference of 80cm and an age of 300 years

□ 良村镇杨将村杨将组 老人说：他的父亲的爷爷栽种的
Yangjiang Group, Yangjiang Village, Liangcun Town An old man said, it was planted by his great grandfather

□ 良村镇杨将村
Yangjiang Village, Liangcun Town
□ 树高 8m 胸围 110cm 树龄 500 年
With a height of 8 meters, a circumference of 110cm and an age of 500 years

□ 良村镇羊化村中心组（茶树湾）
(Chashuwan)Zhongxin Group, Yanghua Village, Liangcun Town
树高 7m ／胸围 90cm 树龄 300 年
With a height of 7 meters, a circumference of 90cm and an age of 300 years.

□ 良村镇杨将村
Yangjiang Village, Liangcun Town
□ 树高 8m 胸围 100cm 树龄 400 年
With a height of 8 meters, a circumference of 100cm and an age of 400 years

□ 良村镇茶园村菜村组
Caicun Group, Chayuan Village, Liangcun Town
□ 树高 8m 胸围 150cm 树龄 900 年
With a height of 8 meters, a circumference of 150cm and an age of 900 years

□ 良村镇羊化村三台山组
Santaishan Group, Yanghua Village, Liangcun Town

□ 树高 6m 胸围 80cm 树龄 300 年
With a height of 6 meters, a circumference of 80cm and an age of 300 years

□ 良村镇羊化村上高原组（福星台）
(Fuxingtai)Shanggaoyuan Group, Yanghua Village, Liangcun Town
□ 树高 6m 胸围 200cm 树龄 1400 年
With a height of 6 meters, a circumference of 200cm and an age of 1400 years

□ 良村镇杨将村杨将组
Yangjiang Group, Yangjiang Village, Liangcun Town
□ 树高 9m 胸围 85cm 树龄 300 年
With a height of 9 meters, a circumference of 85cm and an age of 300 years

□ 良村镇杨将村杨将组
Yangjiang Group, Yangjiang Village, Liangcun Town
□ 树高 10m 胸围 140cm 树龄 800 年
With a height of 10 meters, a circumference of 140cm and an age of 800 years

□ 土城红军渡
Tucheng Red Army Ferry

□ 土城镇长坝村　金秋茶树
Tea trees in the golden autumn, Changba Village, Tucheng Town

□ 土城镇长坝村山上的古茶树
Ancient tea trees on the mountains of Changba Village, Tucheng Town

□ 土城镇同一村三组（大弯子）
(Dawanzi) Sanzu Group, Tongyi Village, Tucheng Town

□ 树高 6m 胸围 80cm 树龄 300 年
With a height of 6 meters, a circumference of 80cm and an age of 300 years

□ 土城镇长坝村二组（双合田）
(Shuanghetian) Erzu Group, Changba Village, Tucheng Town
□ 树高 6m 胸围 80cm 树龄 300 年
With a height of 6 meters, a circumference of 80cm and an age of 300 years

□ 土城镇长坝村二组（双合田）
(Shuanghetian) Erzu Group, Changba Village, Tucheng Town
□ 树高 6m 胸围 120cm 树龄 300 年
With a height of 6 meters, a circumference of 120cm and an age of 300 years

□ 土城镇长坝村二组（双合田）
(Shuanghetian) Erzu Group, Changba Village, Tucheng Town

□ 树高 6m 胸围 170cm 树龄 1100 年
With a height of 6 meters, a circumference of 170cm and an age of 1100 years

□ 土城镇长坝村二组（双合田）
(Shuanghetian) Erzu Group, Changba Village, Tucheng Town
□ 树高 6m 胸围 100cm 树龄 300 年
With a height of 6 meters, a circumference of 100cm and an age of 300 years

☐ 土城镇长坝村二组（双合田）
(Shuanghetian) Erzu Group, Changba Village, Tucheng Town
☐ 树高 6m 胸围 120cm 树龄 300 年
With a height of 6 meters, a circumference of 120cm and an age of 300 years

□ 土城镇同一村三组（大弯子）
(Dawanzi) Sanzu Group, Tongyi Village, Tucheng Town

□ 树高 6m 胸围 60cm 树龄 150 年
With a height of 6 meters, a circumference of 60cm and an age of 150 years

□ 土城镇小坝村中统坝
Zhongtongba, Xiaoba Village, Tucheng Town

□ 土城镇长坝村二组（双合田）
(Shuanghetian) Erzu Group, Changba Village, Tucheng Town
□ 树高 8m 胸围 120cm 树龄 600 年
With a height of 8 meters, a circumference of 120cm and an age of 600 years

□ 土城镇同一村三组（大弯子）古茶树与美景
Ancient tea trees and beautiful scenery, (Dawanzi) Sanzu Group, Tongyi Village, Tucheng Town

□ 双龙乡大坝村唐村坝组（油桂垭）
(Youguiya) Tangcunba Group, Daba Village, Shuanglong Township
□ 树高 10m 胸围 170cm 树龄 1100 年
With a height of 10 meters, a circumference of 170cm and an age of 1100 years

□ 双龙乡双龙村河坝组（新寨）
(Xinzhai) Heba Group, Shuanglong Village, Shuanglong Township
□ 树高 5m 胸围 155cm 树龄 900 年
With a height of 5 meters, a circumference of 155cm and an age of 900 years

双龙乡双龙村河坝组（新寨）
(Xinzhai) Heba Group, Shuanglong Village, Shuanglong Township
树高 8m 胸围 150cm 树龄 900 年
With a height of 8 meters, a circumference of 150cm and an age of 900 years

□ 双龙乡双龙村河坝组（新寨）
(Xinzhai) Heba Group, Shuanglong Village, Shuanglong Township
□ 树高 7m 胸围 100cm 树龄 400
With a height of 7 meters, a circumference of 100cm and an age of 400 years

□ **双龙乡双龙村河坝组（新寨）**
(Xinzhai) Heba Group, Shuanglong Village, Shuanglong Township
□ 树高 8m 胸围 150cm 树龄 900 年
With a height of 8 meters, a circumference of 150cm and an age of 900 years

□ **双龙乡双龙村河坝组（新寨）**
(Xinzhai) Heba Group, Shuanglong Village, Shuanglong Township
□ 树高 8m 胸围 110cm 树龄 500 年
With a height of 8 meters, a circumference of 110cm and an age of 500 years

岩龙乡双龙村河坝组（岩山坪）
(Yanshanping) Heba Group, Shuanglong Village, Shuanglong Township

□ 树高 8m 胸围 124cm 树龄 600 年
With a height of 8 meters, a circumference of 124cm and an age of 600 years

□ 双龙乡双龙村河坝组（岩山坪）
(Yanshanping) Heba Group, Shuanglong Village, Shuanglong Township
□ 树高 6m 胸围 95cm 树龄 300 年
With a height of 6 meters, a circumference of 95cm and an age of 300 years

双龙乡大坝村唐村坝组（油桂垭）
(Youguiya) Tangcunba Group, Daba Village, Shuanglong Township
树高 8m　胸围 90cm　树龄 300 年
With a height of 8 meters, a circumference of 90cm and an age of 300 years

□ 双龙乡双龙村河坪组（大坟林）
(Dafenlin) Heping Group, Shuanglong Village, Shuanglong Township
□ 树高 8m 胸围 120cm 树龄 600 年
With a height of 8 meters, a circumference of 120cm and an age of 600 years

□ 双龙乡双龙村河坪组（水井坎）
(Shuijingkan) Heping Group, Shuanglong Village, Shuanglong Township
□ 树高 8m 胸围 110cm 树龄 500 年
With a height of 8 meters, a circumference of 110cm and an age of 500 years

□ 双龙乡双龙村河坝组（新寨）
(Xinzhai) Heba Group, Shuanglong Village, Shuanglong Township
□ 树高 8m 胸围 110cm 树龄 500 年
With a height of 8 meters, a circumference of 110cm and an age of 500 years

□ 双龙乡双龙村河坝组（新寨）
(Xinzhai) Heba Group, Shuanglong Village, Shuanglong Township
□ 树高 5m 胸围 155cm 树龄 900 年
With a height of 5 meters, a circumference of 155cm and an age of 900 years

□ **永安镇小岗村四坪组**
Siping Group, Xiaogang Village, Yong'an Town

□ 树高 5m 胸围 140cm 树龄 800 年
With a height of 5 meters, a circumference of 140cm and an age of 800 years

□ 永安镇小岗村四坪组
Siping Group, Xiaogang Village, Yong'an Town
Ⓝ 树高 5m 胸围 70cm 树龄 250 年
With a height of 5 meters, a circumference of 70cm and an age of 250 years

永安镇小岗村四坪组
Siping Group, Xiaogang Village, Yong'an Town
树树高 7m　胸围 110cm　树龄 500 年
With a height of 7 meters, a circumference of 110cm and an age of 500 years

□ 东皇街道白坭村
Baini Village, Donghuang Sub-district
□ 树高 9m 胸围 110cm 树龄 500 年
With a height of 8 meters, a circumference of 200cm and an age of 1400 years

□ 东皇街道白坭村
Baini Village, Donghuang Sub-district
□ 树高 9m　胸围 110cm　树龄 500 年
With a height of 9 meters, a circumference of 110cm and an age of 500 years

□ 东皇街道白坭村
Baini Village, Donghuang Sub-district

□ 树高 9m 胸围 105cm 树龄 400 年
With a height of 9 meters, a circumference of 105cm and an age of 400 years

□ 东皇街道白坭村
Baini Village, Donghuang Sub-district

□ 树高 10m 胸围 110cm 树龄 400 年
With a height of 10 meters, a circumference of 110cm and an age of 400 years

□ 东皇街道白坭村
Baini Village, Donghuang Sub-district
☑ 树高 8m 胸围 90cm 树龄 300 年
With a height of 8 meters, a circumference of 90cm and an age of 300 years

□ 东皇街道大陆村付家岩组
Fujiayan Group, Dalu Village, Donghuang Sub-district
□ 树高 6m 胸围 80cm 树龄 300 年
With a height of 6 meters, a circumference of 80cm and an age of 300 years

□ 东皇街道大陆村付家岩组
Fujiayan Group, Dalu Village, Donghuang Sub-district
□ 树高 6m 胸围 90cm 树龄 300 年
With a height of 6 meters, a circumference of 90cm and an age of 300 years

□ 东皇街道大陆村内口组
Fujiayan Group, Dalu Village, Donghuang Sub-district
□ 树高 5m 胸围 80cm 树龄 300 年
With a height of 5 meters, a circumference of 80cm and an age of 300 years

□ 东皇街道大陆村内口组
Fujiayan Group, Dalu Village, Donghuang Sub-district
□ 树高 5m 胸围 150cm 树龄 900 年
With a height of 5 meters, a circumference of 150cm and an age of 900 years

□ 东皇街道大陆村内口组
Neikou Group, Dalu Village, Donghuang Sub-district
□ 树高 5m 胸围 180cm 树龄 1200 年
With a height of 5 meters, a circumference of 180cm and an age of 1200 years

□ 东皇街道大陆村内口组
Neikou Group, Dalu Village, Donghuang Sub-district
□ 树高 7m 胸围 90cm 树龄 300 年
With a height of 7 meters, a circumference of 90cm and an age of 300 years

□ 东皇街道大陆村内口组
Neikou Group, Dalu Village, Donghuang Sub-district
□ 树高 8m 胸围 100cm 树龄 400 年
With a height of 8 meters, a circumference of 100cm and an age of 400 years

□ 东皇街道大陆村内口组
Neikou Group, Dalu Village, Donghuang Sub-district
□ 树高 8m 胸围 120cm 树龄 600 年
With a height of 8 meters, a circumference of 120cm and an age of 600 years

□ 东皇街道大陆村内口组
Neikou Group, Dalu Village, Donghuang Sub-district

□ 树高 9m 胸围 160cm 树龄 1000 年
With a height of 9 meters, a circumference of 160cm and an age of 1000 years

□ 东皇街道大陆村内口组
Neikou Group, Dalu Village, Donghuang Sub-district

□ 树高 8m 胸围 120cm 树龄 600 年
With a height of 8 meters, a circumference of 120cm and an age of 600 years

东皇街道大陆村内口组
Neikou Group, Dalu Village, Donghuang Sub-district
树高 8m 胸围 120cm 树龄 550 年
With a height of 8 meters, a circumference of 120cm and an age of 550 years

东皇街道大陆村付家岩组
Fujiayan Group, Dalu Village, Donghuang Sub-district
树高 9m 胸围 100cm 树龄 400 年
With a height of 9 meters, a circumference of 100cm and an age of 400 years

□ 东皇街道大陆村内口组
Neikou Group, Dalu Village, Donghuang Sub-district
□ 树高 8m 胸围 120cm 树龄 600 年
With a height of 8 meters, a circumference of 120cm and an age of 600 years

□ 九龙街道马蝗坝村新房子组
Xinfangzi Group, Mahuangba Village, Jiulong Sub-district
□ 树高 8m 胸围 80cm 树龄 300 年
With a height of 8 meters, a circumference of 80cm and an age of 300 years

□ 九龙街道马蝗坝村沙地组
Shadi Group, Mahuangba Village, Jiulong Sub-district
□ 树高 8m 胸围 85cm 树龄 300 年
With a height of 8 meters, a circumference of 85cm and an age of 300 years

□ 三岔河镇狮子村 连片的古茶树
Shizi Village, Sanchahe Town, a stretch of ancient tea trees

□ 三岔河镇大坝河村和平组（三岔田）
(Sanchatian) Heping Group, Dabahe Village, Sanchahe Town
□ 树树高 7m 胸围 110cm 树龄 500 年
With a height of 7 meters, a circumference of 110cm and an age of 500 years

三岔河镇大坝村丹溪溪沟组（丫口）
(Yakou) Shanxigou Group, Dabang Village, Sanchahe Town
□ 树高 12m │ 胸围 130cm │ 树龄 700年
With a height of 12 meters, a circumference of 130cm and an age of 700 years

□ 三岔河镇大坝河村和平组（园和坝）
(Yuanheba) Heping Group, Dabahe Village, Sanchahe Town

□ 树高 7m 胸围 160cm 树龄 1000 年
With a height of 7 meters, a circumference of 160cm and an age of 1000 years

□ 三岔河镇狮子村后庄组（弯弯头）
(Wanwantou) Houzhuang Group, Shizi Village, Sanchahe Town

□ 树高 15m 胸围 170cm 树龄 1100 年（老鹰茶）
With a height of 15 meters, a circumference of 170cm and an age of 1100 years (Eagle Tea)

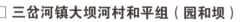

□ 三岔河镇大坝河村和平组（三岔田）
(Sanchatian) Heping Group, Dabahe Village, Sanchahe Town
□ 树高 6m 胸围 150cm 树龄 900 年
With a height of 6 meters, a circumference of 150cm and an age of 900 years

□ 三岔河镇大坝河村和平组（三岔田）
(Sanchatian) Heping Group, Dabahe Village, Sanchahe Town
□ 树高 5m 胸围 90cm 树龄 300 年
With a height of 5 meters, a circumference of 90cm and an age of 300 years

□ 三岔河镇大坝河村和平组（三岔田）
(Sanchatian) Heping Group, Dabahe Village, Sanchahe Town
□ 树高 6m 胸围 150cm 树龄 900 年
With a height of 6 meters, a circumference of 150cm and an age of 900 years

□ 三岔河镇狮子村
Shizi Village, Sanchahe Town
□ 树高 9m，胸围 90cm，树龄 300 年
With a height of 9 meters, a circumference of 90cm and an age of 300 years

□ 三岔河镇狮子村后庄组（弯弯头）
(Wanwantou) Houzhuang Group, Shizi Village, Sanchahe Town
□ 树高 6m，胸围 85cm，树龄 300 年
With a height of 6 meters, a circumference of 85cm and an age of 300 years

□ 大山深处 ——九里十三弯
Thirteen curves along a route of nine miles, in the depth of the mountains

□ **大坡镇飞鸽村群利组（木林台）**
(Mulintai) Qunli Group, Feige Village, Dapo Town
□ 树高 15m 胸围 175cm 树龄 1100 年（古属茶）
With a height of 15 meters, a circumference of 175cm and an age of 1100 years (Eagle Tea)

□ 大坡镇黄河坝村田坪组
Tianping Group, Huangheba Village, Dapo Town
□ 树高 9m 胸围 85cm 树龄 300 年（老鹰茶）
With a height of 9 meters, a circumference of 85cm and an age of 300 years

□ **大坡镇黄河坝村田坪组**
Tianping Group, Huangheba Village, Dapo Town
□ 树高 15m 胸围 110cm 树龄 500 年（老鹰茶）
With a height of 15 meters, a circumference of 110cm and an age of 500 years (Eagle Tea)

□ **大坡镇黄河坝村田坪组**
Tianping Group, Huangheba Village, Dapo Town
□ 树高 15m 胸围 115cm 树龄 500 年（老鹰茶）
With a height of 15 meters, a circumference of 115cm and an age of 500 years (Eagle Tea)

大坡镇野茶树
Wild tea tree Dapo Town

□ 大坡镇野茶树
Wild tea tree Dapo Town

□ **大坡镇野茶树**
Wild tea tree Dapo Town

□ 大坡镇野茶树
Wild tea tree Dapo Town

□ 大坡镇野茶树
Wild tea tree Dapo Town

□ 寨坝镇永胜村永兴组（家长湾）
(Jiazhangwan) Yongxing Group, Yongsheng Village, Zhaiba Town
□ 树高 10m 胸围 190cm 树龄 1300 年
With a height of 10 meters, a circumference of 190cm and an age of 1300 years

□ 寨坝镇永胜村双龙组（大屋基）
(Dawuji) Shuanglong Group, Yongsheng Village, Zhaiba Town

□ 树高 8m，胸围 110cm 树龄 400 年
　With a height of 8 meters, a circumference of 110cm and an age of 400 years

□ 寨坝镇永胜村双龙组（大屋基）
(Dawuji) Shuanglong Group, Yongsheng Village, Zhaiba Town
□ 树高 8m　胸围 80cm　树龄 300 年
With a height of 8 meters, a circumference of 80cm and an age of 300 years

□ 马临街道沔山村十组（水井门口）古茶树群
Ancient tea tree clusters, (Shuijingmenkou) Shizu Group, Mianshan Village, Malin Sub-district

马临街道沔山村十组（水井门口）
(Shuijingmenkou) Shizu Group, Mianshan Village, Malin Sub-district

树高 7m 胸围 160cm 树龄 1000 年
With a height of 7 meters, a circumference of 160cm and an age of 1000 years

马临街道沔山村十组（水井门口）
(Shuijingmenkou) Shizu Group, Mianshan Village,Malin Sub-district

遮天蔽日大茶树叶
Sky-sheltering big tea tree leaves

□ 马临街道沔山村十组（水井门口）
(Shuijingmenkou) Shizu Group, Mianshan Village, Malin Sub-district

□ 树高 7m 胸围 130cm 树龄 700 年
With a height of 7 meters, a circumference of 130cm and an age of 700 years

□ 马临街道沔山村十组（水井门口）
(Shuijingmenkou) Shizu Group, Mianshan Village, Malin Sub-district
□ 树高 7m 胸围 160cm 树龄 1000 年
With a height of 7 meters, a circumference of 160cm and an age of 1000 years

□ 马临街道沔山村十组（水井门口）
(Shuijingmenkou) Shizu Group, Mianshan Village,Malin Sub-district
□ 树高 7m 胸围 130cm 树龄 700 年
With a height of 7 meters, a circumference of 130cm and an age of 700 years

□ 马临街道沔山村十组（水井门口）
(Shuijingmenkou) Shizu Group, Mianshan Village,Malin Sub-district
□ 树高 7m 胸围 160cm 树龄 1000 年
With a height of 7 meters, a circumference of 160cm and an age of 1000 years

□ 官店镇黄桃村太平组（七瓦池）
(Qiwachi) Taiping Group, Huangtao Village, Guandian Town
□ 树高 9m，胸围 100cm，树龄 400 年
With a height of 9 meters, a circumference of 100cm and an age of 400 years

□ 官店镇黄桃村杨平组（李家）
(Lijia) Yangping Group, Huangtao Village, Guandian Town
□ 树高 8m　胸围 150cm　树龄 900 年
With a height of 8 meters, a circumference of 150cm and an age of 900 years

□ 官店镇黄桃村杨平组（七瓦池）
(Qiwachi) Taiping Group, Huangtao Village, Guandian Town
□ 树高 6m 胸围 80cm 树龄 300 年
With a height of 6 meters, a circumference of 80cm and an age of 300 years

□ 官店镇黄桃村太平组（七瓦池）
(Qiwachi) Taiping Group, Huangtao Village, Guandian Town
□ 树高 8m 胸围 80cm 树龄 300 年
With a height of 8 meters, a circumference of 80cm and an age of 300 years

官店镇黄桃村杨平组（李家）
(Lijia) Yangping Group, Huangtao Village, Guandian Town
树高 10m 胸围 180cm 树龄 1200 年
With a height of 10 meters, a circumference of 180cm and an age of 1200 years

□ 官店镇黄桃村杨平组（李家）
(Lijia) Yangping Group, Huangtao Village, Guandian Town
□ 树高 8m　胸围 170cm　树龄 1100 年
With a height of 8 meters, a circumference of 170cm and an age of 1100 years

□ 仙源镇金桥村南山组（猴子岩）
(Monkey Rocks) Nanshan Group, Jinqiao Village, Xianyuan Town
□ 树高 7m　胸围 90cm　树龄 300 年
With a height of 7 meters, a circumference of 90cm and an age of 300 years

□ 仙源镇金桥村大坪组　一片古茶树
A stretch of ancient tea trees, Daping Group, Jinqiao Village, Xianyuan Town

□ **官店镇黄桃村太平组（七瓦池）**
(Qiwachi) Taiping Group, Huangtao Village, Guandian Town
□ 树高 7m 胸围 80cm 树龄 300 年
With a height of 7 meters, a circumference of 80cm and an age of 300 years

□ 桃林镇龙凤村龙凤组（高角元）
(Gaojiaoyuan) Longfeng Group, Longfeng Village, Taolin Town
□ 树高 7m 胸围 127cm 树龄 600 年
With a height of 7 meters, a circumference of 127cm and an age of 600 years

□ 桃林镇天龙村高坪组（沟洞门）茂盛古树叶
Lush ancient tree leaves, (Goudongmen) Gaoping Group, Tianlong Village, Taolin Town

□ 古茶树枝
Ancient tea tree branches

□ **桃林镇天龙村高坪组（高角元）**
(Gaojiaoyuan) Gaoping Group, Tianlong Village, Taolin Town
□ 树高 6m 胸围 127cm 树龄 600 年
With a height of 6 meters, a circumference of 127cm and an age of 600 years

□ **桃林镇天龙村高坪组（沟洞门）**
(Goudongmen) Gaoping Group, Tianlong Village, Taolin Town
□ 树高 9m 胸围 114cm 树龄 500 年
With a height of 9 meters, a circumference of 114cm and an age of 500 years

□ 桃林镇天龙村高坪组（王角房子）
(Wangjiaofangzi) Gaoping Group, Tianlong Village, Taolin Town
□ 树高 6m 胸围 90cm 树龄 300 年
With a height of 6 meters, a circumference of 90cm and an age of 300 years

□ **桃林镇龙凤村龙凤组（余家）**
(Yujia) Longfeng Group, Longfeng Village, Taolin Town

□ 树高 8m 胸围 175cm 树龄 1100 年
With a height of 8 meters, a circumference of 175cm and an age of 1100 years

□ **桃林镇龙凤村龙凤组（寨子）**
(Zhaizi) Longfeng Group, Longfeng Village, Taolin Town

□ 树高 6m 胸围 130cm 树龄 700 年
With a height of 6 meters, a circumference of 130cm and an age of 700 years

□ 桃林镇天龙村高坪组〔王角房子〕
(Wangjiaofangzi) Gaoping Group, Tianlong Village, Taolin Town
□ 树高 6m 胸围 116cm 树龄 500 年
With a height of 6 meters, a circumference of 116cm and an age of 500 years

桃林镇兴隆村麻坪组（金宝台）
(Jinbaotai), Maping Group, Xinglong Village, Taolin Town
树高 13m 胸围 220cm 树龄 1800 年
With a height of 13 meters, a circumference of 220cm and an age of 1800 years

□ **桃林镇天龙村高坪组（王角房子）**
(Wangjiaofangzi) Gaoping Group, Tianlong Village, Taolin Town
□ 树高 7m 胸围 106cm 树龄 400 年
With a height of 7 meters, a circumference of 106cm and an age of 400 years

□ **桃林镇天龙村高坪组（王角房子）**
(Wangjiaofangzi) Gaoping Group, Tianlong Village, Taolin Town
□ 树高 7m 胸围 112cm 树龄 500 年
With a height of 7 meters, a circumference of 112cm and an age of 500 years

□ 桃林镇天龙村高坪组　古树茶果
Ancient tree tea and fruits, Gaoping Group, Tianlong Village, Taolin Town

□ **桃林镇兴隆村麻坪组（金宝台）**
(Jinbaotai), Maping Group, Xinglong Village, Taolin Town
□ 树高 8m 胸围 120cm 树龄 600 年
With a height of 8 meters, a circumference of 120cm and an age of 600 years

□ **桃林镇兴隆村麻坪组（金宝台）**
(Jinbaotai), Maping Group, Xinglong Village, Taolin Town
□ 树高 9m 胸围 110cm 树龄 500 年
With a height of 9 meters, a circumference of 110cm and an age of 500 years

☐ **桃林镇天龙村高坪组（王角房子）**
(Wangjiaofangzi) Gaoping Group, Tianlong Village, Taolin Town

☐ 树高 7m 胸围 130cm 树龄 700
With a height of 7 meters, a circumference of 130cm and an age of 700 years

☐ **桃林镇兴隆村麻坪组（金宝台）**
(Jinbaotai), Maping Group, Xinglong Village, Taolin Town

☐ 树高 12m 胸围 120cm 树龄 600 年
With a height of 12 meters, a circumference of 120cm and an age of 600 years

□ 桃林镇天龙村高坪组（高角元）
(Gaojiaoyuan) Gaoping Group, Tianlong Village, Taolin Town
□ 树高 6m　胸围 100cm　树龄 500 年
With a height of 6 meters, a circumference of 100cm and an age of 500 years

桃林镇兴隆村黄村组（落石溪）
(Luoshixi) Huangcun Group, Xinglong Village, Taolin Town
树高 8m　胸围 115cm　树龄 600 年
With a height of 8 meters, a circumference of 115cm and an age of 600 years

桃林镇兴隆村麻坪组（金宝台）
(Jinbaotai), Maping Group, Xinglong Village, Taolin Town
树高 10m　胸围 126cm　树龄 600 年
With a height of 10 meters, a circumference of 126cm and an age of 600 years

□ 桃林镇天龙村高坪组（高角元）
□ 树高 7m 胸围 140cm 树龄 800 年
(Gaojiaoyuan) Gaoping Group, Tianlong Village, Taolin Town
With a height of 7 meters, a circumference of 140cm and an age of 800 years

□ 桃林镇龙凤村龙凤组（余家）
(Yujia) Longfeng Group, Longfeng Village, Taolin Town
□ 树高 8m 胸围 120cm 树龄 600 年
With a height of 8 meters, a circumference of 120cm and an age of 600 years

□ 桃林镇龙凤村龙凤组（大湾头）
(Dawantou) Longfeng Group, Longfeng Village, Taolin Town
□ 树高 7m 胸围 230cm 树龄 2000 年
With a height of 7 meters, a circumference of 230cm and an age of 2000 years

□ 温水镇目里村联合组 古茶树林
Ancient tea forest, Lianhe Group, Muli Village, Wenshui Town

□ **温水镇大水村油沙组（板干上）**
(Banganshang)Yousha Group, Dashui Village, Wenshui Town
□ 树高 8m 胸围 90cm 树龄 300 年
With a height of 8 meters, a circumference of 90cm and an age of 300 years

□ **温水镇大水村油沙组（板干上）**
(Banganshang)Yousha Group, Dashui Village, Wenshui Town
□ 树高 10m 胸围 130cm 树龄 700 年
With a height of 10 meters, a circumference of 130cm and an age of 700 years

温水镇目里村联合组
Lianhe Group, Muli Village, Wenshui Town

树高 8m　胸围 150cm　树龄 900 年
With a height of 8 meters, a circumference of 150cm and an age of 900 years

□ 温水镇目里村联合组（沙子龙）
(Shazilong)Lianhe Group, Muli Village, Wenshui Town
□ 树高 8m 胸围 80cm 树龄 300 年
With a height of 8 meters, a circumference of 80cm and an age of 300 years

□ 温水镇目里村联合组（小溪沟）
(Xiaoxigou)Lianhe Group, Muli Village, Wenshui Town
□ 树高 9m 胸围 114cm 树龄 500 年
With a height of 9 meters, a circumference of 114cm and an age of 500 years

□ **温水镇目里村联合组（小溪沟）**
(Xiaoxigou)Lianhe Group, Muli Village, Wenshui Town
□ 树高 7m 胸围 80cm 树龄 300 年
With a height of 7 meters, a circumference of 80cm and an age of 300 years

□ 温水镇大水村油沙组（板干上）
(Banganshang)Yousha Group, Dashui Village, Wenshui Town

□ 树高 8m 胸围 80cm 树龄 300 年
With a height of 8 meters, a circumference of 80cm and an age of 300 years

□ 温水镇大水村油沙组（板干上）
(Banganshang)Yousha Group, Dashui Village, Wenshui Town

□ 树高 7m 胸围 80cm 树龄 300 年
With a height of 8 meters, a circumference of 80cm and an age of 300 years

□ 温水镇大水村油沙组（板干上）
(Banganshang)Yousha Group, Dashui Village, Wenshui Town
□ 树高 7m 胸围 140cm 树龄 800 年
With a height of 7 meters, a circumference of 140cm and an age of 800 years

□ 温水镇炉村河坝组（簸箕岩）
(Bojiyan) Heba Group, Lucun Village, Wenshui Town

□ 树高 7m 胸围 80cm 树龄 300 年
With a height of 7 meters, a circumference of 80cm and an age of 300 years

□ 温水镇目里村回龙组（草房头）
(Caofangtou) Huilong Group, Muli Village, Wenshui Town

□ 树高 8m 胸围 90cm 树龄 300 年
With a height of 8 meters, a circumference of 90cm and an age of 300 years

□ **温水镇大水村油沙组（板干上）**
(Banganshang)Yousha Group, Dashui Village, Wenshui Town
□ 树高 7m　胸围 90cm　树龄 300 年
With a height of 7 meters, a circumference of 90cm and an age of 300 years.

□ 温水镇目里村联合组 房前屋后的古茶树
Ancient tea trees surrounding the houses, Lianhe Group, Muli Village, Wenshui Town

□ 温水镇炉村河坝组（簸箕岩）
(Bojiyan) Heba Group, Lucun Village, Wenshui Town
□ 树高 8m 胸围 110cm 树龄 500 年
With a height of 8 meters, a circumference of 110cm and an age of 500 years

□ 温水镇目里村回龙组（草房头）
(Caofangtou) Huilong Group, Muli Village, Wenshui Town
□ 树高 9m 胸围 150cm 树龄 600 年
With a height of 9 meters, a circumference of 150cm and an age of 600 years

□ **温水镇炉村河坝组（簸箕岩）**
(Bojiyan) Heba Group, Lucun Village, Wenshui Town
□ 树高 8m 胸围 100cm 树龄 500 年
With a height of 8 meters, a circumference of 100cm and an age of 500 years

□ **温水镇大水村油沙组（板干上）**
(Banganshang)Yousha Group, Dashui Village, Wenshui Town
□ 树高 7m　胸围 120cm　树龄 600 年
With a height of 7 meters, a circumference of 120cm and an age of 600 years

□ **温水镇大水村油沙组（板干上）**
(Banganshang)Yousha Group, Dashui Village, Wenshui Town
□ 树高 12m　胸围 100cm　树龄 400 年
With a height of 12 meters, a circumference of 100cm and an age of 400 years

□ **温水镇目里村联合组（小溪沟）**
(Xiaoxigou)Lianhe Group, Muli Village, Wenshui Town
□ 树高 9m 胸围 130cm 树龄 700 年
With a height of 9 meters, a circumference of 130cm and an age of 700 years

温水镇目里村联合组田边的大茶树
Big tea trees aside the fields, Lianhe Group, Muli Village, Wenshui Town

□ **温水镇大水村油沙组（板干上）**
(Banganshang)Yousha Group, Dashui Village, Wenshui Town
□ 树高 12m　胸围 150cm　树龄 900 年
With a height of 12 meters, a circumference of 150cm and an age of 900 years

□ 温水镇目里村同心组（三合头）
(Sanhetou) Tongxin Group, Muli Village, Wenshui Town
□ 树高 10m 胸围 110cm 树龄 500 年
With a height of 10 meters, a circumference of 110cm and an age of 500 years

□ 温水镇桥头村先锋组（金榜田）
(Jinbangtian) Xianfeng Group, Qiaotou Village, Wenshui Town
□ 树高 14m 胸围 85cm 树龄 300 年
With a height of 14 meters, a circumference of 85cm and an age of 300 years

□ 温水镇目里村联合组
Lianhe Group, Muli Village, Wenshui Town
□ 树高 11m 胸围 120cm 树龄 600 年
With a height of 11 meters, a circumference of 120cm and an age of 600 years

□ 温水镇目里村联合组（小溪沟）
(Xiaoxigou)Lianhe Group, Muli Village, Wenshui Town
□ 树高 9m 胸围 130cm 树龄 700 年
With a height of 9 meters, a circumference of 130cm and an age of 700 years

□ **温水镇目里村联合组（小溪沟）**
(Xiaoxigou)Lianhe Group, Muli Village, Wenshui Town
□ 树高 9m 胸围 200cm 树龄 1400 年
With a height of 9 meters, a circumference of 200cm and an age of 1400 years

□ **温水镇目里村同心组（白乳田）**
(Bairutian) Tongxin Group, Muli Village, Wenshui Town
□ 树高 8m 胸围 85cm 树龄 300 年
With a height of 8 meters, a circumference of 85cm and an age of 300 years

□ 古树嫩芽
Tender sprouts on ancient trees

□ 温水镇大水村半坡组（上寨）
(Shangzhai) Banpo Group, Dashui Village, Wenshui Town
□ 树高 10m 胸围 93cm 树龄 300 年
With a height of 10 meters, a circumference of 93cm and an age of 300 years

□ **温水镇大水村龙洞组（曾木垭）**
(Zengmuya) Longdong Group, Dashui Village, Wenshui Town
□ 树高 8m 胸围 110cm 树龄 500 年
With a height of 8 meters, a circumference of 110cm and an age of 500 years

□ 温水镇大水村半坡组（上寨）
(Shangzhai) Banpo Group, Dashui Village, Wenshui Town
□ 树高 8m 胸围 103cm 树龄 400 年
With a height of 8 meters, a circumference of 103cm and an age of 400 years

□ 温水镇目里村回龙组（草房头）
(Caofangtou) Huilong Group, Muli Village, Wenshui Town
□ 树高 7m 胸围 90cm 树龄 300 年
With a height of 7 meters, a circumference of 90cm and an age of 300 years

□ 温水镇大水村半坡组（上寨）
(Shangzhai) Banpo Group, Dashui Village, Wenshui Town

□ 树高 10m　胸围 140cm　树龄 800 年
With a height of 10 meters, a circumference of 140cm and an age of 800 years

温水镇炉村河坝组（簸箕岩）
(Bojiyan) Heba Group, Lucun Village, Wenshui Town
□ 树高 9m 胸围 110cm 树龄 500 年
With a height of 9 meters, a circumference of 110cm and an age of 500 years

□ **温水镇目里村回龙组**
Huilong Group, Muli Village, Wenshui Town
□ 树高 10m 胸围 250cm 树龄 2500 年
With a height of 10 meters, a circumference of 250cm and an age of 2500 years

□ 茶果磊磊
Plentiful tea leaves and fruits

新的时代
A New Era

特别感谢

贵州省林业科学研究院
贵州大学农学院

习水县领导：

向承强　陈　钊　陈永清　王仲勇　冯俊峰
安家兴　宋聚宗　沈建通　李建生　苟明利　袁仁端　邱会儒　冯沛红　黄定旭　严古军
刘华江　胡　伟　陈昌海　田　勇　刘维清　田　莉　龚小松　苏方涛　张　蒙

习水县有关部门、乡镇、企业及个人：

政府办　宣传部　公安局　经贸局　文广局　文　联　住建局　财政局　旅游局　农牧局
水务局　林业局　城管局　统计局　招商局　投资局　扶贫办　城建局　县委党校
东皇街道　马临街道　杉王街道　九龙街道　土城镇　温水镇　永安镇　双龙乡
仙源镇　良村镇　同民镇　桑木镇　寨坝镇　桃林镇　二里镇　民化镇　程寨镇
官店镇　大坡镇　三岔河镇

习水宾馆　宋窖博物馆　三溪集团　习水县企业家商会

陈平国　袁晓宁　涂卫东　赵红光　冯　斌　黄　敏　贺　焱　刘自力　刘锦定　谢　永　冯宗尧　郑银辉　李朝福　张世其　陆　位　周彭林
袁　浩　袁　令　康昭平　郑银行　曹　波　李学文　袁祥忠　周江海　何华勤　张明理　何世文　任皓皓　李庆忠　陈长辉　陆　勇　穆文付
丁宗才　任　波　汤中海　邓忠常　文建业　肖敏敏　陈小松　孙县章　冯安炯　罗永杰　张元铨　赵　晋　刘　弦　何　平　周　玲　袁继勇
陈　飞　刘　峰　穆小峰　江　远　罗　华　汪骞文　镱　璇　穆仁启　涂利高　杨金勇　杨　俊　黄文涛　赵　军　李红平　袁仲秋　李　良
许翰林　陈明杰　罗　杰　李寿伦　罗　力　项　杰　曹廷昌　雷宏钟　冯　彻　刘　坤　王正梅　袁　化　何　龙　涂旭东　杨铁军　马江红
陈　胜　马剑飞　王　靖　娄必华　杨　胜　杨正刚　肖江波　周朝林　罗其虎　余亭廷　吴开别　穆伦禄　阮雨飞　何华勤　刘圣勇　朱仁祥
冯文琼　李　彬　闫跃华　何　勇　朱大昌　朱建波　游先萍　赵小刚　罗　洪　周　羽

村　民：

穆世坤　陈应民　任栋杉　王　永　菜大义　袁国超　冯文强　冯文德　袁万方　谢帝强　谢学宽　李星国
穆贤泽　郑银生　冷洪明　冷洪权　任章书　何永辉　王运强　王运钟　王德财　张永宪　任灯明　邹群英
陆昌志　唐光明　戴天平　饶连邦　袁天智　王　芳　谢永贵　刘光树　袁智刚　石国凡　杨青林　杨伦权
杨伦远　杨忠明　袁　勇　王章龙　王忠财　罗其元　王德尧　王忠贵　陶德开　袁天智　代天平　代青培
李学文　袁守青　陈江文　何永祥　黄无权　余　佳　陈晓华

总 协 调：王志雄
保　　障：邱继军　张　伶　邹德生　曹　维　吕咏梅
工作人员：周枞胜　马星禄　杨　近　王腾跃　赵云波　康永骑　袁　敏　张先国
策　　划：北京百年传奇文化发展有限公司　　贵州鳛国故里茶业有限公司

摄　　影：刘太华　刘廷明　应智群　周东亚　高 立
责任编辑：杨美艳
装帧设计：侯东华　袁 泉

图书在版编目（CIP）数据

习水古茶树 / 谭智勇 主编．-- 北京 : 人民出版社，
2018.4
　ISBN 978-7-01-019113-3

　Ⅰ．①习… Ⅱ．①谭… Ⅲ．①茶树－习水县－摄影集
Ⅳ．① S571.1-64

中国版本图书馆 CIP 数据核字 (2018) 第 054977 号

习 水 古 茶 树
XISHUI GU CHASHU

主编 谭智勇

人 民 出 版 社 出版发行

（100706　北京市东城区隆福寺街 99 号）

北京诚顺达印刷有限公司印刷　　新华书店经销

2018 年 4 月第 1 版　　2018 年 4 月北京第 1 次印刷
开 本：889 毫米 ×1194 毫米 1/12　印张：24
字数：8 千字　印数：1—2000 册

ISBN 978-7-01-019113-3　　定价：1600.00 元

邮购地址 100706　北京市东城区隆福寺街 99 号
人民东方图书销售中心　电话（010）65250042　65289539